- **Effect of consecutive primes in creating composite numbers:**

- **History of prime numbers:**

- **Number N is a prime number if it is greater than 1 and is divisible only by one and itself; otherwise, it is a composite number.**

- **We have a fundamental theorem of arithmetic:**
  Every composite number can be written as a product of prime numbers uniquely.
  Primes are building blocks of all numbers.

- **We have The Sieve of Eratosthenes:**
  It gave a method to generate all prime numbers between 1 and a given defined number.

- We have a prime number theorem
  $\pi(N) = N/\ln(N)$.
  it gives an approximate count for the primes within the number N.
- In this paper or research, we have a set of consecutive primes that do not include primes 2 and 5 starting with prime 3 and the last prime is F and we want to know how many composite numbers are created by each of those consecutive primes.

My theorem statement:

Definitions:

Array PTBP

- **It is the following Array of odd numbers**

$$\begin{vmatrix} 1 & 3 & 7 & 9 \\ 11 & 13 & 17 & 19 \\ 21 & 23 & 27 & 29 \\ 31 & 33 & 37 & 39 \\ 41 & 43 & 47 & 49 \\ 51 & 53 & 57 & 59 \end{vmatrix}$$

And so on....

- **For a given set of consecutive primes whose numbers =n that start with prime number 3 and end with prime number F and not including prime number 2 and prime number 5**

I.e.

Set=[3,7,11,13,..............................., F]

S=product of those consecutive primes

$$S = \prod_{i=3}^{i=F} (i)$$

i= consecutive values of prime numbers 3, 7, 11, 13,......., F

Range=R_k = 10 $\times$ S $\times$ k =$2 \times 5 \times s \times K$

Where k = [1, 2, 3, 4, ...,
∞(infinity)

- i.e.

  $R_1 = 10 \times S \times 1$

  $R_2 = 10 \times S \times 2$

  And so on

- Number of composite numbers that belong to Array PTBP and created by the effect of those consecutive primes within the range R_k

$$= [(K \times 4^{\times \frac{S}{3}}) + ($$

$$\sum_{j=7}^{j=F} (K \times 4 \times (\frac{S}{j}) \times$$

$i = prime\ number\ before\ current\ prime\ number\ j$

$$\prod_{i=7} \qquad (\frac{i-1}{i})$$

$)]-$

(n)

Where j =consecutive values of prime numbers

7, 11, 13,.............., F.

And i= consecutive values of prime numbers

3, 7, 11, 13,........, prime before current j prime.

- **The previous formula can be applied for any number of consecutive primes that start with prime 3**

$$(k \times 4 \times \frac{S}{3})$$

- **The first term represents the count of unique**

Composite numbers +1 that belong to the Array PTBP and are created by prime number 3 within the range

$$R_k = 10 \times S \times k$$

- **The second term**

$$\sum_{j=7}^{j=F} (K \times 4 \times (\frac{S}{j}) \times$$

$i = $ *prime number befor current prime number j*

$$\prod_{i=7} \qquad (\frac{i-1}{i})$$

Represent the count of unique Composite numbers+n-1 that belong to the Array PTBP and are created by each prime number after the prime number 3 within the range $R_k = 10 \times S \times k$

- **The third term (-n)**

Subtracting n (number of consecutive primes starting from prime number 3) because the count of numbers generated from those consecutive primes includes the count of those primes in the range $R_k = 10 \times S^{\times} k$

- **<u>Explanation and proof for my theory:</u>**
  The philosophy of the solution to reach an understanding of the role of the prime numbers in the formation of composite numbers is to reach something common between the numbers that belongs to the Array PTBP.
- It is easy to conclude that any prime number(except for prime number 2 and number 5) must end with one of those numbers (1,3,7,9) since any number with the last digit(2,4,8,0,5) cannot be a prime number (except prime number 2 and number 5), since it is divisible by 2 or 5.
- So the probability that the number is prime number is contained in the Array

PTBP (except prime number 2 and number 5).

- If we multiply any prime number p with the numbers within m row in the array PTBP.

  Where the first row is (1, 3, 7, 9   for m=1.

  The result will equal to ((p x (m-1) x 10) + (p x 1) , (p x (m-1) x 10) + (p x 3) , (p x (m-1) x 10) + (p x 7) , (p x (m-1) x 10) + (p x 9)).

- And it is It's obvious that the first part of the four results [(p x (m-1) x 10)] depends on the m value and equal to [0, p x 10, p x 20, p x 30,......] for m=[1,2,3,4.......]

- The second part is fixed 4 values [p x 1, p x 3, p x 7, p x 9]   that must belong to Array PTBP because the one's place of any prime number that belongs to

Array PTBP =[1 or 3 or 7 or 9] and hence the multiplication result must belong to the Array PTBP

- so we can define a new concept (number cycles) as repeated slices (each slice length = p x 10)in which the multiples of each prime number p that belong to the set of consecutive primes exist at fixed 4 ratios relative to each slice .

- So we can use the number cycle concept to understand the behavior of consecutive primes in creating composite numbers.

- We can unify number cycles for any number of consecutive primes that belong to that Array PTBP by multiplying those consecutive primes and then multiplying the result by ten so that each number of those consecutive primes has completed a full cycles

- i.e.

S=product of those primes

$$S = \prod_{i=3}^{i=F} (i)$$

- the multiples of number 3 will cancel the probability of four numbers in this Array PTBP to be a prime number in each number cycle after the first number cycle for number three since 3 is a prime number in the first number cycle, and only 3 probabilities in the first cycle are canceled.
- The same for prime number 7 and the other prime numbers in the set of consecutive primes.

- The prime number 3 takes priority, followed by the prime 7, then prime 11, and so on
- i.e. 21 counts for prime 3 not for prime 7 and so on
- The problem of interference in the effect of prime numbers in creating composite numbers is solved by the reduction factor in the formula.

Which =

$$i = prime\ number\ before\ current\ prime\ number\ j$$
$$\prod_{i=7} \left(\frac{i-1}{i}\right)$$

This will be explained later.

- Now we will explain each term in the formula

$$\left[\left(K \times 4^{\times \frac{S}{3}}\right) + \left(\sum_{j=7}^{j=F} \left(K \times 4 \times \left(\frac{S}{j}\right) \times \prod_{\substack{i=7 \\ i=\text{prime number befor current prime number } j}}^{} \left(\frac{i-1}{i}\right)\right)\right)\right] -$$

(n)

$$\left(k \times 4 \times \frac{S}{3}\right)$$

- **The first term** **represents the count of unique Composite numbers +1 that belong to the Array PTBP and are created by prime number 3 within the range $R_k = 10 \times s \times k$.**

This term can be explained as the count of number cycles of prime number 3 which has a length

= 3 x 10 = 30 and each cycle have 4 unique counts of divisibility of the prime number 3

- And as mentioned above the prime number 3 takes priority.

- **The second term**

$$\sum_{j=7}^{j=F} (K \times 4 \times (\frac{S}{j}) \times$$

$i = $ *prime number before current prime number j*

$$\prod_{i=7} \qquad \left(\frac{i-1}{i}\right)$$

The first part of this term $(k \times 4 \times (S/j)$

Represent the count of unique Composite numbers +n-1 that belong to Array PTBP and are created by each j prime number after the prime 3 within the range $R_k = 10 \times S \times k$

But it needs a reduction factor which is

$$\prod_{i=7}^{i = \text{prime number before current prime number } j} \left(\frac{i - 1}{i}\right)$$

- **This adjusts the result due to the effect of previous prime numbers before the current prime because prime number 3 takes priority in creating composite numbers then prime 7 then prime 11 and so on**

- **NOW WE WILL MAKE A GENERAL PROOF FOR THE REDUCTION FACTOR:**

  Suppose we have a prime number that belongs to the Array PTBP.

  This Array has very special characteristics and facts that will be explained later.

  Take any column of this Array

Every three rows have one value that is divisible by 3

- Every 7 rows have one value that is divisible by 7
- And so on and we will prove it
- Suppose that prime number =p (with ones place must =1 or =3 or =7 or =9)
- if we multiply prime p with the elements in the m row in PTBP
- The result will equal to ((p x (m-1) x 10) + (p x 1) , (p x (m-1) x 10) + (p x 3) , (p x (m-1) x 10) + (p x 7) , (p x (m-1) x 10) + (p x 9))
- The result is 4 values that belong to this Array PTBP each time we multiply the prime number p with the next m row

- if we multiply the prime number p with the numbers of the first column in PTBP the results will be ((p x (m-1) x 10)+(p x 1) depending on the m value and all results belong to a unique particular column in the array PTBP depending on the one's place of the prime number p (1 or 3 or 7 or 9) and the one's place of the elements of that particular column (1)

- it means that the multiplication results will be uniformly spaced from the initial value p x 1 and the spacing between two consecutive results will be [p x (m-1+1) x 10] – [p x (m-1) x 10]=p x 10 x (m-m+1)=p x 10 x 1=p x 10

- Or we can see that each p row after the initial value has unique divisibility of prime number p

- And similar result is if we multiply prime p with the elements of the second column in PTBP

With different initial values = p x 3

- And similar result is if we multiply prime p with the elements of the third column in PTBP

  With different initial values = p x 7
- And similar result is if we multiply prime p with the elements of the fourth column in PTBP

  With different initial values = p x 9
- Each result belongs to a column of the Array PTBP that column depends on the one place of the prime p (1 or 3 or 7 or 9) and the common one place of the column of Array PTBP (1 for the first column, or 3 for the second column, or 7 for the third column, or 9 for the fourth column)
- The following table shows the ones place or last digit of the multiplication result

(last digit) of prime p	The last digit of the first column of the array PTBP =1	The last digit of the first column of the array PTBP =3	The last digit of the first column of the array PTBP =7	The last digit of the first column of the array PTBP =9
1	1	3	7	9
3	3	9	1	7
7	7	1	9	3
9	9	7	3	1

- The initial values whether it = p x 1 or p x 3 or p x 7 or p x 9 = X depend on the p-value and the value of

integer $\left(\dfrac{X}{10}\right)$ = y then the X value exists at the y+1 row within the array PTBP and the next divisibility of prime number P within each column exists at the y+1+p row within the array PTBP

- It means before the X value there are y rows and the row of the value X =1 row and between the row of the X value and the row of the next divisible value of prime number p there are p-1 row
- We can subtract y from p-1 and it will equal p-1-y
- Then X value exists within y+1+p-1-y rows i.e. exists within the first p rows

- And each of the successive divisibility within the same column exists within the next y+1+p-1-y rows i.e. the p rows
- i.e. Each p   successive row in any column within Array PTBP will have one value that is divisible by p
- Where p is a prime number that belongs to the Array PTBP

- Note: the multiples of any prime greater than 3 must exist in a separate row within the Array PTBP

- Then if we have a set of n consecutive primes that belong to the Array PTBP
- Set=[3,7,11,13,...............Z,p_i,p_{i+1}............
  ....,F]

  And   S=product of those consecutive primes

  i.e.
  $$S = \prod_{i=3}^{i=F} (i)$$

- Then prime number 3 throw-out results affect the prime p_{i+1} throw-out result within Array PTBP
- Count of common divisibility for both prime number 3 and prime number pi_{+1} within range S x 10
  = (4 x S)/ (3 x p_{i+1})
- The corresponding values for this count are represented by part of the Array PTBP with the number of rows C_1=(S)/ (3 x p_{i+1}) with values =

$3 \times p_{i+1} \times 1$, $3 \times p_{i+1} \times 3$, $3 \times p_{i+1} \times 7$, $3 \times p_{i+1} \times 9$

$3 \times p_{i+1} \times 11$, $3 \times p_{i+1} \times 13$, $3 \times p_{i+1} \times 17$, $3 \times p_{i+1} \times 19$

And so on

- And prime p_{i+1} has a total count for divisibility within the range 10 x S

  $= 4 \times S / p_{i+1}$

- The corresponding values for this count are represented by part of the Array PTBP with the number of rows $C_2 = (S)/(p_{i+1})$ with values =

  $p_{i+1} \times 1$, $p_{i+1} \times 3$, $p_{i+1} \times 7$, $p_{i+1} \times 9$

  $p_{i+1} \times 11$, $p_{i+1} \times 13$, $p_{i+1} \times 17$, $p_{i+1} \times 19$

  And so on

- So reduction factor = $1-[((4 \times S)/ (3 \times p_{i+1}))/(4 \times S/ p_{i+1})]=1-(1/3)=2/3$

- And prime p_{i+1} has total unique counts of divisibility within range 10 x S after excluding divisibility of prime 3 $= (4 \times S/ p_{i+1}) \times (2/3)$

- Then prime 7 throw-out results affect the prime p_{i+1} throw-out result within Array PTBP
- Number of common divisibility for both prime number 7 and prime number p_{i+1} within range S x 10
  = $(4 \times S)/ (7 \times p_{i+1})$
- And the number of common divisibility for both prime 7 and prime p_{i+1} within range S x 10
  = $[(4 \times S)/ (7 \times p_{i+1})] \times (2/3)$ after excluding divisibility of prime 3
- Prime p_{i+1} has total counts for divisibility within the range 10 x S
  = $(4 \times S/ p_{i+1}) \times (2/3)$ after excluding divisibility of prime number 3

- So reduction factor due to the effect of prime number 7 on unique divisibility on prime number p_{i+1}= $1-[((4 \times S)/(7 \times$

$p_{i+1}))\times(2/3)/((4 \times S/ p_{i+1}) \times (2/3))]=1-(1/7)=6/7$

- And prime p_{i+1} has total unique counts of divisibility within the range 10 x S after excluding divisibility of prime number 3 and prime number 7= (4 x S/ p_{i+1}) x (2/3)x(6/7)

- Let

  $T_i=$

  reduction factor due to the effect of prime i

- And the number of common divisibility for both prime p_i and prime p_{i+1} within range S x 10

  $= [(4 \times S)/ (p_i \times p_{i+1})] \times \prod_{i=3}^{i=Z} T_i$

  After excluding the divisibility of primes 3, 7,..........., Z

- Let $B=\prod_{i=3}^{i=z} T_i$

- Prime p_{i+1} has total counts for divisibility within the range 10 x S

  $= (4 \times S/ p_{i+1}) \times B$

After excluding divisibility of primes 3, 7,.........., Z

So reduction factor due to the effect of prime p_i on the unique divisibility of prime p_{i+1} = $1-[((4 \times S)/ (p_i \times p_{i+1})) \times (B)/((4 \times S/ p_{i+1}) \times (B))]=1-(1/p_i)=(p_i-1)/p_i$

And if $p_i=V$

Prime p_{i+1} has total unique counts of divisibility within range $10 \times S$ after excluding divisibility of primes 3, 7,.........., Z,p_i

$$= (4 \times S/ p_{i+1}) \times \prod_{i=3}^{i=V} [(i-1)/i]$$

- Which explains the reduction factor within the formula?

$i =$ prime number before current prime number j

$$\prod_{i=7} \left(\frac{i-1}{i}\right)$$

Randomness in the prime numbers

- There is no randomness in the prime numbers

- The effect of prime numbers [3,7] with K=1

We can apply the formula for any number of consecutive primes

S= 3×7 = **21**

Range R$_1$=10 $\times 21 \times 1$= **210**

Effect of number three in the first unified cycle including prime number three

$$4 \times \frac{21}{3} = 28$$

- Effect of prime number 7 in the first unified cycle including primer number seven

$$= 4 \times \frac{21}{7} \times \frac{2}{3} = 8$$

As we see the effect of the prime in creating composite numbers within that defined range reduces according to this amazing formula, the most effective one is prime number three, then prime number seven, and so on, and the total number of composite numbers within the Array PTBP in the first unified cycle due to the effect of those two consecutive primes =36-2=34 composite numbers

And =36 composite numbers in the other unified cycles

- The effect of prime numbers [3, 7, 11, 13] with k =1
  We can apply a formula for any number of consecutive primes
  S= $3 \times 7 \times 11 \times 13$ = **3003**

Range R1=10 $\times 3003 \times 1$ **= 30030**

Effect of number three in the first unified cycle including prime three

$$4 \times \frac{3003}{3} = 4004$$

- **Effect of prime number 7 in the first unified cycle including prime number seven**

$$= \ ^{4 \times \frac{3003}{7} \times \frac{2}{3}} = \textbf{1144}$$

Effect of prime number 11 in the first unified cycle including prime number eleven

$$= \ ^{4 \times \frac{3003}{11} \times \frac{2}{3} \times \frac{6}{7}} = \textbf{624}$$

- **Effect of number 13 in the first unified cycle including prime thirteen**

$$= \ ^{4 \times \frac{3003}{13} \times \frac{2}{3} \times \frac{6}{7} \times \frac{10}{11}} = \textbf{480}$$

As we see the effect of the prime number in creating composite numbers within that defined range reduces

according to this amazing formula, the most effective one is prime number three, then prime number seven, and so on, and the total count of composite numbers that belong to Array PTBP in the first unified cycle due to the effect of those four consecutive primes =6252-4=6248 composite numbers

- And =6252 composite numbers in the other unified cycles

- So my research solves the problem of the patterns formed within the distribution of prime numbers, those patterns of composite numbers (that are repeated regularly each unified cycle after the first unified cycle since the first unified cycle contains those consecutive primes) occur because the behavior of those consecutive primes in creating composite numbers

- **The result of the mentioned proved formula is independently checked by Python programming language and it is accurate 100% without error or approximations.**

www.ingramcontent.com/pod-product-compliance
Lightning Source LLC
Chambersburg PA
CBHW050528290526
45786CB00007B/2738